MW01308996

The Scoop on Ice Cream

Other books by Vicki Cobb

Chemically Active! Experiments You Can Do at Home
Making Sense of Money
The Monsters Who Died: A Mystery about Dinosaurs
Science Experiments You Can Eat
Magic . . . Naturally: Science Entertainments and Amusements
More Science Experiments You Can Eat
Bet You Can!: Science Possibilities to Fool You (with Kathy Darling)
Bet You Can't!: Science Possibilities to Fool You (with Kathy Darling)
The Secret Life of Cosmetics
The Secret Life of School Supplies
How to Really Fool Yourself: Illusions for All Your Senses
The Secret Life of Hardware: A Science Experiment Book
Lots of Rot
Fuzz Does It!
Gobs of Goo
How the Doctor Knows You're Fine
Brave in the Attempt: The Special Olympics Experience
Supersuits
Sneakers Meet Your Feet

The Scoop on Ice Cream

by Vicki Cobb

illustrated by G. Brian Karas

Little, Brown and Company

Boston Toronto

Second Printing

Library of Congress Cataloging in Publication Data
Cobb, Vicki.
The scoop on ice cream.

(How the world works series)
Summary: Outlines the ingredients and making of ice cream and the role played by the manufacturers, retailers, and suppliers of this popular dessert. Includes a taste test and recipe for the homemade variety.
1. Ice cream, ices, etc. — Juvenile literature.
[1. Ice cream, ices, etc.] I. Karas, G. Brian, ill.
II. Title. III. Series.
TX795.C63 1985 637.4 85-6881
ISBN 0-316-14895-4

BP

Published simultaneously in Canada
by Little, Brown & Company (Canada) Limited

Printed in the United States of America

This series is dedicated to Louis Sarlin,
the teacher who gave me the best year of my childhood
and the key to my place in the world.

Contents

1. A Cold, Sweet Story — 3
2. Milk and Cream — 6
3. Sugar — 13
4. Vanilla and Chocolate — 21
5. The Ice-Cream Factory — 33

An Ice-Cream Taste Test — 46

Make Your Own Ice Cream — 50

Index — 55

The author gratefully acknowledges the help of the following people: Dawn Brydon of the Milk Industry Foundation, Matt Whitehead of Welsh Farms, Diane McKuen of the Chocolate Manufacturers Association, Steve Scarangella and Russ Davies of Refined Sugar, Professor Dick H. Kleyn of Rutgers University, Nicholas Kominus of U.S. Cane Sugar Refiners Association, John LeSauvage of Schrafft's, Margaret, Christopher, and Kent Johnson of Heinchon's Dairy, Carl P. Hetzel of Virginia Dare Extract Co., Inc., Ted Van Leer of Van Leer Chocolate Corp., and Norma Harrison of The Marcus Group, Inc.

1.
A Cold, Sweet Story

Bet you can make the juices run in your mouth. Think about digging into a giant scoop of your favorite ice cream. You put the tip of your spoon into the ice cream and press down. The ice cream cuts easily. You put the mound of ice cream on the spoon in your mouth and hold it on your tongue. It tastes sweet and creamy. It feels cold and smooth. Slowly it gets smaller as your warm mouth makes it melt. Melted ice cream coats your tongue. The roof of your mouth feels cool. You keep making little swallows of melted ice cream. Yum! Are your juices running yet?

Now I'll bet you feel like having some ice cream. No problem. There are lots of places you can get it. Maybe there's some in your freezer. If not, you can buy some easily at the nearest grocery or supermarket. Before you can eat ice cream you must own it. That means either you pay for it or the person who bought it gives it to you. Ice-cream eaters are the last people to own ice cream. But they are not the first. And they are often not the second. Different people owned the ice cream ingredients and the ice cream itself before you became the last ice cream owner.

Your scoop of ice cream has a story to tell. It's a story of where it begins. Ice cream is made of milk and cream, sugar and flavoring such as vanilla, chocolate, or strawberry. It may contain candy or nuts or fruit. It's a story of how these ingredients get to the ice-cream factory or plant. Some of them come across oceans. It's a story of how the ingredients are mixed and frozen in the ice-cream plant. And it's a story about money. Money from the sale of ice cream pays the people who make it and the people who supply the ingredients. Many people earn a living making the ice cream you buy. It's a story of one way the world works. This book tells that story.

Want to know the scoop on ice cream? Read on.

2. Milk and Cream

The milk we buy in the supermarket is liquid food. It is made in the bodies of cows. A cow makes milk to feed her baby calf. A cow is a big animal. She can produce a lot of milk, much more than she needs to feed one calf. So cows are raised on farms to supply milk for drinking and for making other foods like cheese and ice cream.

The people who raise cows and sell their milk are *dairy farmers.* For about two months a year a dairy cow does not give milk. These are the last two months before she has her calf. Each dairy cow has one calf every year. After her calf is born her body starts making milk. The farmer milks the cow. The more milk he takes from her body, the more milk her body makes, up to a point. Amazing but true!

A cow turns food and water into milk. She eats over 41 pounds of food a day. This food is made of grains and seeds, hay, chopped corn, and grass. She also drinks between 10 and 20 gallons of water. There are 160 glasses of water in 10 gallons. So it's no surprise that a cow spends most of her day eating and drinking. Out of all this food and water, her body makes about 20 quarts of milk a day. In America, there's one cow for every 14 people. Imagine. One cow provides enough milk to feed 14 people with milk, cheese, and ice cream every day!

A cow's milk collects in her *udder,* or milk bag, that hangs just in front of her hind legs. A calf gets milk from its mother by sucking on one of the four *teats* that hang from the udder. A farmer can milk a cow by pulling on her teats. Warm milk squirts out into a pail. Years ago, farmers had to milk their cows by hand. An expert milker could empty a cow in about 20 minutes. But now machines do all the milking. Farmers feed their calves milk in buckets. Most of the milk is sold. Here's how a dairy farmer supplies the ice-cream maker with milk.

The farmer's day starts very early. The first milking takes place about five o'clock in the morning. The second milking is late in the afternoon. A group of cows steps into the milking parlor. There is a separate

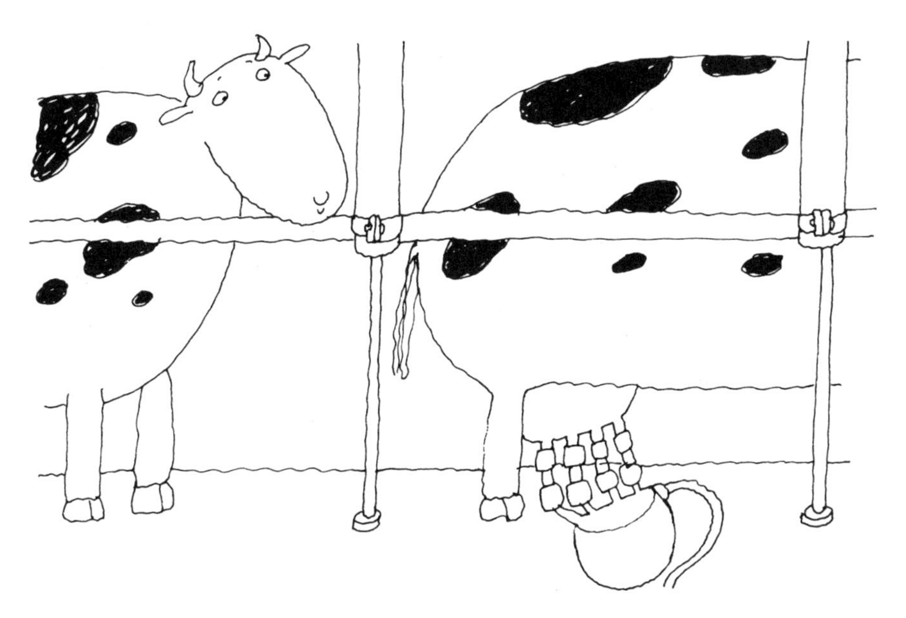

stall for each cow. The farmer washes off the teats and udder of each cow to make sure they are clean. He attaches a tube of the milking machine to each teat. The milking machine sucks the milk out of the udder. In three minutes, the udder is empty. The milking machines are removed and the cows go back to doing what they do most of the day: namely, eating. The next group of cows steps into the milking parlor.

The milk goes straight from the cow's udder through the tube into a pipe. It joins with the milk from other cows. The pipe delivers the milk to a large tank outside the milking parlor. Milk that comes straight from the cow is *raw milk.*

Calves and people are not the only living things that like milk as food. There are tiny plants, called *bacteria,* that you can see only with a microscope. Bacteria are floating in the air. If they land on the milk, they will live in it and use it as food. Their wastes will go into the milk. The milk will spoil. Since raw milk travels directly to the storage tank through pipes, it is exposed to very little air. There is little chance for bacteria to land.

When the milk comes from the cow, it is warm from the cow's body. Bacteria grow especially well in warm milk. Bacteria don't grow as quickly if the milk is cold. So the storage tank for raw milk is like a refrigerator. It makes the milk cold. But even cold milk will spoil if it has enough time. So milk must be used quickly. The raw milk stays in the farmer's tank for two days, at most. Usually it is collected every day.

Milk is mostly water. But it also contains fat and solids called *proteins.* If you let fresh raw milk stand, the part containing most of the fat rises to the top. The fat is *butterfat* and the liquid that contains butterfat is *cream*. An electric stirrer in the raw-milk tank keeps the cream from separating. The tank is made of stainless steel that can be kept very clean. People who work with milk, from dairy farmers to milk packaging workers to ice-cream factory workers, must wash all the tanks and pipes that hold milk every day.

Ice-cream makers are dairy farm customers. They buy raw milk. Most ice-cream plants are located near dairy farms. The plants send tank trucks to pick up the raw milk from the farms. A tank truck holds about

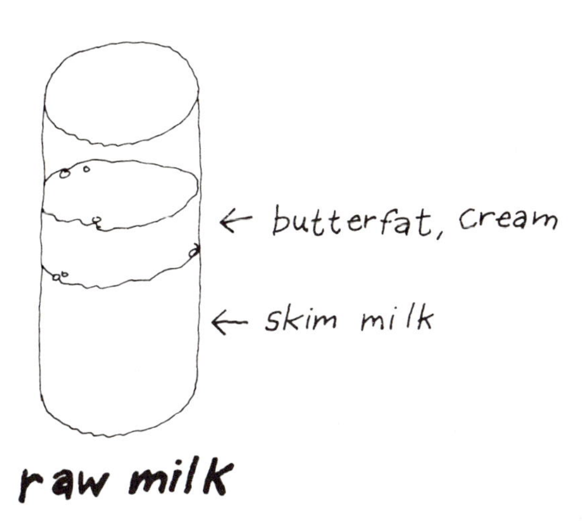

4,000 gallons of milk. The milk tanker may stop at more than one farm to fill up. You've seen milk tankers on the road. A milk tanker is usually silver. It is like a thermos bottle lying on its side. It keeps the cold raw milk cold.

Ice-cream factories need fresh milk every day. The delivery is on its way.

3.
Sugar

There is a special kind of grass called *sugarcane* that grows on many large farms in hot parts of the world. It's a tall grass, as grasses go. And the stalk is thick. When it is fully grown, the stalk is between one and two stories high. You could not circle most of the fully grown stalks with your thumb and middle finger. It's very hard work to cut this grass down because it is tough and woody. In some places, men cut the stalks with large knives. On more modern farms, machines do the work.

Why do people raise this tough, tall grass? If you chewed on a stalk you would know why. It is very sweet. Sugarcane and another plant, sugar beet, are the main sources of sugar for your sugar bowl and, of course, for making ice cream.

Sugarcane leaves make sugar out of air and water. All green plants can do this. Sugar is used by plants as food. Plants store food until it is needed. Most plants change sugar into starch to store it. (Flour is a kind of starch.) Starch isn't sweet. But sugarcane doesn't change sugar into starch. It stores sugar instead of starch in its stalks, or *canes.*

The sugar that's in the canes is not useful. After the cane has been harvested, the problem is to get the sugar out. No small job. The first step takes place fairly close to the sugarcane farm. Cut sugarcane is rushed by truck or railroad to a nearby mill. The sugar in the canes starts changing into less valuable sugar as soon as the cane is cut. So it is very important to start getting it out quickly.

At the mill the cane is chopped and then passed through huge rollers that crush the juice from the cane. Juice pours from the rollers into tanks. During the harvest, which lasts between 3 and 6 months, the mills work constantly, 24 hours a day, 7 days a week. A strong chemical is heated with the juice to remove *impurities,* substances that are not sugar. The impurities include dirt from the fields and pieces of the plant. The impurities settle to the bottom as a kind of "mud." Then the sugary liquid is boiled in a special way.

When you boil something on the stove, it is hot. But sugar burns easily and can't be boiled at such a high temperature without taking the chance that it will be ruined. So sugarcane juice is boiled in a space

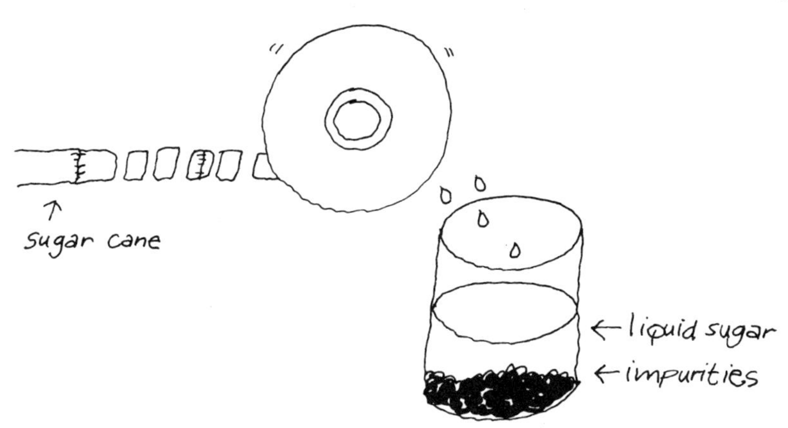

where air has been pumped out. Such a space is called a *vacuum.* The vacuum lets the sugar boil at a temperature that is cooler than the one for boiling on a stove. The sugar is now protected from burning.

As the water from the juice boils off, sugar crystals start to form. These sugar crystals are not like the sugar in your sugar bowl. They are coated with a brown, sticky substance called *molasses.* This brown coated sugar, called *raw sugar,* is not ready for use in your sugar bowl or for making ice cream. There are still many impurities in it. But it is ready to leave the area where it grew. It is ready to be loaded onto a bulk freighter and make the trip, perhaps across the ocean, to the factory where it will become white and pure, the *sugar refinery.*

Most sugar refineries are on water deep enough for large oceangoing freighters, such as the Great Lakes or the Hudson River. The freighter ties up at the refinery's dock. The dock has large cranes. At the end of each crane is a *clam* that drops into the cargo bay and scoops up the raw sugar. The clam empties the sugar into funnel-like hoppers that deliver the sugar

to moving belts which pour the sugar into a large warehouse. Yellowjackets and bees swarm around the dock, feeding on spilled sugar. They know a free lunch when they see one!

The warehouse is huge. It holds a mountain of raw sugar that looks just like sand. But it doesn't smell like sand. Sniff your sugarbowl or, better yet, brown sugar, and you'll know what it smells like. In the warehouse the raw sugar is weighed and then stored until it is refined.

A sugar refinery is a big factory, several blocks long. The workers wear hard hats to protect their heads. It's easy to bump your head on the miles of pipes in the refinery. Here's what happens to the raw sugar as a refinery prepares it.

- Raw sugar is mixed with a watery sugar syrup in a *mingler.* Mingling softens the molasses coating each sugar crystal. The raw sugar crystals don't dissolve in the syrup that washes them.
- Washed sugar crystals are put in a spinning tank called a *centrifuge.* The centrifuge acts like the spin cycle in a washing machine. Water carrying molasses spins away from the sugar crystals.
- Next, the washed raw sugar crystals are dissolved in

water. This makes a syrup. Now the sugar can flow through pipes.

– The syrup moves to a tank where substances are added to remove still more impurities. Some impurities sink to the bottom as dark brown "mud." Others float to the top, where they can be skimmed off.

– Now the sugar syrup, which is light brown in color, can be strained by passing through filters. These filters are screens that let the syrup through but trap larger impurities that are not dissolved in water.

– The syrup is still light brown. The color is caused by impurities that are small enough to pass through the screen filters. But there is a substance that can collect the brown impurities. It's finely ground-up charcoal. So the light brown syrup drips through a charcoal filter that is three stories tall. When it comes out the bottom, it is finally colorless and pure. Finished sugar syrups are called *liquors.*

– The liquors are boiled in a vacuum. Some of them will be piped to areas where crystals will form. These crystals will be dried and put in bags for sale in supermarkets and grocery stores. The finished product is *granulated* sugar.

– Another finished product stays in liquid form. It is piped to another part of the refinery, where it will be put in tank cars that are trailers on trucks.

Tank trucks back into the syrup loading and shipping area. The loader climbs on top. He takes off the cap and puts in a funnel. He puts the end of the pipeline into the funnel. Pumps send liquid sugar into the tanks. Most tanks hold 4,000 gallons.

After the tanker has been loaded with liquid sugar, the driver sets off to make his delivery. One customer who buys tankloads of liquid sugar is the ice-cream factory. The shipment is on its way.

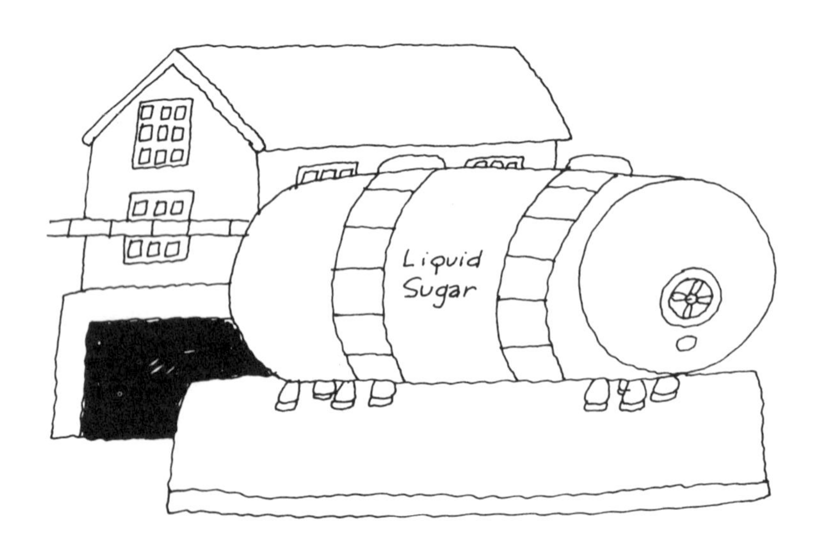

4. Vanilla and Chocolate

Orchids are showy flowers that you often see pinned to someone's shoulder on special occasions. They grow wild in very hot parts of the world. The orchids florists sell are usually grown in greenhouses that create a tropical climate. In tropical parts of the world, in Mexico and on islands off the coast of Africa, there are farms that grow a certain kind of orchid. These special orchids are not grown for their beauty and they will not decorate shoulders. They are grown because they are the source of one of the most popular flavors in the world — namely, vanilla.

Orchids, like all flowers, do an important job for their plant. They produce the seeds that can become new orchid plants. In four to six weeks, a flower can produce six to ten pods full of tiny black seeds. A pod looks like a string bean. It is about 7 inches long and is green with a yellow tip. It is these pods, known as *vanilla beans*, that provide us with vanilla flavoring.

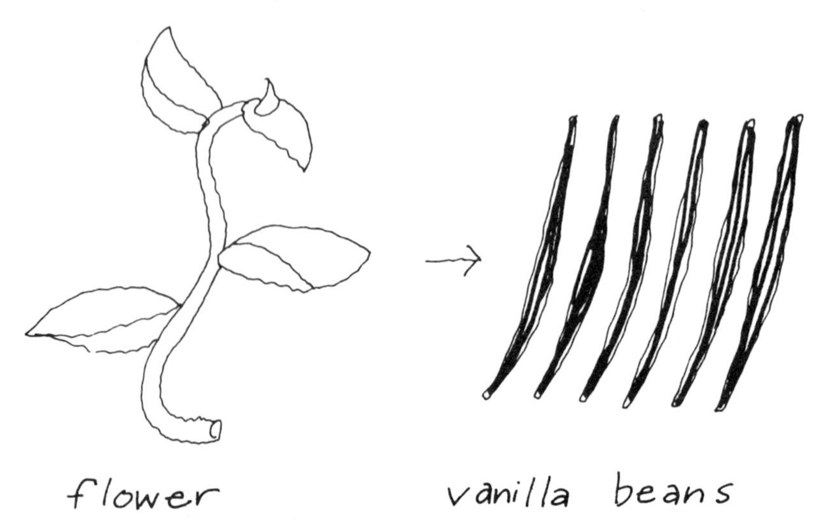

Although vanilla beans are produced in several places, Madagascar produces most of the vanilla beans in the world. Madagascar is a large island off the east coast of southern Africa. Vanilla beans from Madagascar are called *Bourbon* beans.

If you looked at a freshly picked vanilla bean, you would never guess that it was the source of vanilla. It has no vanilla taste or smell. It is filled with a thick liquid that can give you a rash if you rub it on your skin. Before a bean has a vanilla taste it must be *cured* and aged. Curing starts changes in certain substances in the vanilla bean so that its flavor develops. Curing and aging take about nine months.

In Madagascar, beans are picked from November to February. Curing starts when the freshly picked beans are put in hot water for about three minutes. Then they are spread in the hot sun and covered with blankets. The beans start to sweat. Liquid appears on their surface. They sweat all day. At night they are stored in covered bins. Daily sweating and nights of cooling go on for days or weeks, depending on the weather. Finally they are put on long frames in store-

houses where a fresh breeze can cool and dry them. When the curing is finished, the Bourbon beans are dark brown and shiny with oils that smell of vanilla.

Cured vanilla beans are still not ready to be used in ice cream. But they are ready to be shipped to factories that take the vanilla flavoring out of the beans. Vanilla flavoring that has been removed from vanilla beans and can be used by ice-cream makers is called *vanilla extract.* Most of the factories that make vanilla extract for American ice-cream makers are located in America.

When vanilla beans arrive at the flavoring factory you can smell their wonderful aroma. The way people make vanilla extract is something like the way coffee is made in homes. First the vanilla beans are chopped. Chopped vanilla beans are like ground coffee. The chopped beans are put in stainless-steel tanks. A mixture of alcohol and water is *percolated* through the chopped beans. Percolators (including coffee percolators) send the liquid through the beans several times. In the vanilla percolator, the vanilla flavoring dissolves into the liquid. Every time the liquid passes through the beans, more vanilla flavoring dissolves in it. Finally all the flavoring is out of the beans and in the alcohol-water mixture.

If you want to know what vanilla extract smells like, check the spices in your house. There's a good chance you'll find a bottle. Smell the bottle of vanilla extract. Heaven! But don't taste it. It tastes terrible. Vanilla tastes good only when a small amount is mixed into cake batter or ice cream.

The vanilla extract is now stored in stainless-steel or wooden drums and aged for several months. Aged vanilla extract is then poured into plastic gallon jugs

or in 55-gallon steel drums for delivery to an ice-cream factory or to some other customer. Vanilla extract keeps for a long time without spoiling. The ice-cream plant may get a shipment once a month or every few months depending on how fast it is used up. Vanilla extract is a very strong flavor. A little goes a long way.

Like vanilla, chocolate also comes from warm parts of the world. And like vanilla, chocolate comes from a seedpod. The source of chocolate is the *cacao* (ka-KAH-oh) tree, an evergreen with large glossy leaves. Cacao farms are in tropical parts of the world where

the trees have warm weather and a lot of rain. About 20 different countries in Africa and the Americas grow most of the crop that becomes chocolate.

Cacao trees sprout clusters of waxy white or pink flowers on their trunks and older branches. These flowers produce melon-shaped pods that must be cut off the trees by hand. Ripe pods, the size of grapefruits, can be found on cacao trees all year long, but there is usually one main harvest and another smaller harvest.

The workers bring the pods to an area near the

fields. The pods are piled up. Then the pods are split open with one or two blows from a long-handled knife. Inside are 20 to 50 cream-colored beans. They are scooped out and the husk is thrown away. The air makes their color change to light purple. These cacao beans will become cocoa and chocolate. But they are a long way from that now.

The cacao beans are put into boxes or baskets and allowed to *ferment* for several days. Fermenting starts the process that changes substances inside the fleshy beans to chocolate. Nothing has to be done to the beans to get them to ferment. It just happens. Fermenting beans give off a sour acid smell as well

as heat. The temperature may go as high as 125 degrees, which kills the center of the bean and further develops the substances that will become chocolate. The fully fermented bean is a rich brown without a purple center. It is ready to be dried and put in sacks for shipping.

The next stop is a factory that turns cacao beans into chocolate for candy and, of course, for ice cream. Although there are some chocolate factories in the United States, many of the world's chocolate factories are in Holland. American ice-cream makers often buy chocolate from Dutch companies. But no matter where a chocolate factory is, all chocolate factories have some fantastic heavy machinery that treats millions of cacao beans. It takes about 400 beans to make one pound of chocolate.

The cacao beans arrive at the chocolate factory in sacks weighing between 100 and 200 pounds. The first thing that happens to the beans is a good cleaning. Bits of dried pulp and pieces of pod that are in with the beans are removed, along with other dirt from the fields. Next comes the step that turns the dried beans into chocolate. They are roasted. Roasting is done in huge, stainless-steel tube-like ovens. The ovens turn constantly, making sure the beans are roasted evenly. If you stood near one of these roasting ovens, you would smell chocolate. Mmmmm!

Roasting also makes the shells on the beans brittle and dry. That makes it easy to do the next step, which removes the shells. The beans pass between two rollers that crack the shells. Fans blow away the cracked shells. The shelled beans are called *nibs.* Nibs are pure chocolate. But they are still not quite ready to be used for candy or ice cream.

The nibs move on to mills where they are ground up between large grinding stones or heavy steel disks. The grinding produces heat. The warm, crushed nibs become a river of dark brown chocolate called *chocolate liquor.* When this chocolate liquid is cooled in molds, it is sold as baker's chocolate in grocery stores for use in making homemade cakes and cookies. Chocolate liquor is bitter. It is not yet good to eat.

Most ice-cream makers do not use chocolate li-

quor. They use powdered cocoa. Cocoa is made from chocolate liquor. Here's how. The chocolate liquor is pumped into giant presses that will put pressure on the hot liquid. Up to 25 tons of pressure is put on the chocolate liquor. This pressure separates the liquor into two parts. One part is a yellow fat known as *cocoa butter.* Cocoa butter is used to make chocolate candy bars, among other things. After the cocoa butter is removed, all that is left is a light brown cake. This cake is then ground into powder and put in bags. It is cocoa.

Fifty-pound bags of powdered cocoa are loaded onto trucks. Gallons of vanilla extract are loaded onto other trucks. They take to the road. The chocolate and vanilla shipments are on their way to the ice-cream plant.

5.
The Ice-Cream Factory

Workers at the ice-cream plant receive shipments of the ingredients for their product. An ice-cream plant might buy cocoa by the trailerload of 800 50-pound bags. The bags are piled up on wooden trays called *pallets.* Small *forklift trucks* move pallets of cocoa into the dry-storage warehouse.

Another shipment in bags that comes by the truckload is *milk solids.* Milk solids is skim milk that has had the water removed. Milk solids is also known as powdered milk. It makes the ice cream have a richer taste. Pallets of bags of milk solids are also moved to the dry-storage area.

Sugar tankers deliver their loads in their own special area. After the truck arrives, a sample of sugar goes to the ice-cream plant laboratory. It is tested to see how sweet it is and if it contains anything that could spoil the ice cream. If it passes the tests, the load is weighed and then it is unloaded through pipes into a storage tank.

When a milk tanker arrives, the same kind of thing happens. A sample of milk is tested in the lab. It is tested to see that it is fresh. It is also tested to see how much butterfat is in the milk. The amount of butterfat in milk can be different from tankload to tankload. Butterfat gives ice cream its creamy taste. The amount of the other ingredients that will be mixed to make ice cream depends on how much butterfat is present. The laboratory worker calls the milk unloader with the test results. If the milk passes the tests, it is pumped from the tanker through pipes into refrigerated storage tanks. Then the inside of the tanker is washed thoroughly before it is sent on its way.

Some of the raw milk will remain whole. Some of it will be separated into heavy cream and skim milk. The cream is put in another refrigerated storage tank, waiting to be used for ice cream.

All the ingredients are now on hand to start making ice cream. The first step is to make the *ice-cream mix*. An ice-cream plant makes large batches of mix, say 1,000 gallons at one time. The man who makes

the mix is called the *mix master.* It is the mix master's job to make sure that the finished ice-cream mix is always the same. The amount of butterfat in milk and cream can be different from day to day. So the mix master adjusts the amounts of the other ingredients.

There are two kinds of mixes, white and chocolate. One thousand gallons of white mix is made of about 480 gallons of milk, 290 gallons of cream, 500 pounds of milk solids, and 200 gallons of liquid sugar. Chocolate mix is the same as white mix except that 350 pounds of cocoa are added. There is one other ingredient that goes into the mix. It is called a *stabilizer.* A stabilizer is important when the ice cream is frozen. I'll tell you more about stabilizers later. Compared to the amount of the rest of the ingredients, there is very little stabilizer. The mix master adds 15 to 20 pounds.

The mix is made in a stainless-steel tank. The mix master measures out the proper amounts of milk, cream, and liquid sugar by setting dials on meters. The three liquids are pumped from their storage tanks to the mixing tank through pipes. Valves close the pipes when the proper amount of each ingredient has been delivered. So the liquid ingredients are measured out automatically.

The dry ingredients, the milk solids, the stabilizer, and (for chocolate) the cocoa are dumped into the tank by the mix master. A huge beater in the tank thoroughly mixes the ingredients. The mix master makes white ice-cream mix early in the day. He makes chocolate later. All the equipment must be thoroughly washed before he can make white mix again. But that's not a problem. All ice-cream equipment must be washed every day. It's the law.

The law states that all milk and milk products must be treated to kill any harmful bacteria they may contain. Long ago all milk that was sold was raw, straight from the farm. Milkmen delivered the milk door-to-door. There was no refrigeration, so milk had to be used quickly or it would get sour. But sour milk was not the problem. Disease was.

Milk, 100 years ago, sometimes carried bacteria for two diseases that often killed children. One disease was *tuberculosis.* Tuberculosis is a killer whose tiny sores form in the lungs and make it very hard to breathe. The second killer germ in milk caused *undulant fever.* The word "undulant" means "waves." In this disease the fever comes in waves. A child with undulant fever can have a temperature of 105 degrees in the evening. By morning his temperature will be normal. But it rises again as the day goes on. Many children died of undulant fever and tuberculosis from milk.

A French scientist, Louis Pasteur, discovered how to kill disease-causing bacteria more than 125 years ago. He found that when milk is heated to 145 degrees

for 30 minutes, all the germs are killed. Heating milk to kill germs is a process named for its discoverer. It's called *pasteurization*. However, pasteurization does not kill bacteria that make milk sour.

Today, all milk and milk products are pasteurized. Any plant that processes milk or milk products must keep its equipment very clean. Hot detergents are pumped through all the pipes and tanks when mixing is finished every day. The worker who does the washing signs a book. Inspectors from the government come very often. They look at the equipment. They check the books. But they seldom find any problems. It's not good business to produce milk or ice cream that makes the public sick. Ice-cream makers have no problem following the rules.

The milk and cream in ice cream can be pasteurized before it is used in the mix. But it doesn't have to be. Usually the mix is pasteurized after it is mixed. It is pumped into the pasteurizer, where it is heated to about 170 degrees and held at this temperature for about 25 seconds. This is the modern fast method of pasteurization. Then the mix is cooled quickly. The mix master tastes the mix. It has a slightly "cooked" flavor, like warm milk that has cooled. A sample is taken to the lab. There the butterfat and milk solids in the mix are measured to double-check the formula. The mix is now ready for the next step, becoming ice cream.

If you took some ice-cream mix and stuck it in the freezer, you would not get ice cream. You would get a solid block of ice-cream-flavored ice. See for yourself. Let some ice cream melt completely. Then put it back into the freezer. The refrozen ice cream is not very good to eat. Ice-cream mix has to be frozen in a very special way in order to become ice cream. Here's why.

The liquid in ice cream that can cause problems when ice cream is frozen is none other than water. Water is in milk, it's in cream, and it's in liquid sugar. When water freezes, it becomes solid ice crystals. The idea in making ice cream is to make the ice crystals as tiny as possible. Then the frozen water will be spread evenly through the butterfat, milk solids, and sugar. The ice cream will feel smooth to your tongue.

There are two ways to make sure the ice crystals are kept very small. First, freeze the ice cream as quickly as possible. Slow freezing causes large ice crystals. Second, beat the ice cream while it is freezing.

triple barrel freezer

Beating keeps the ice crystals small, but it also adds another important ingredient for the proper feel of ice cream in your mouth. This ingredient is none other than air. Air makes ice cream soft and light. When the proper amount of air is added, one gallon of ice-cream mix becomes about 2 gallons of ice cream.

The ice-cream freezer is a triple-barreled machine. Each barrel freezes and whips the mix at the same time. When the ice cream comes out of the freezer, it is still quite soft. It is just like soft ice cream! No doubt this is where the idea of soft-ice-cream stores came from. Soft ice cream must be made fresh and served right away if it is to be soft. But in the ice-cream factory, the soft ice cream is still not quite finished. Fruits and nuts are added, depending on the flavor. (Flavoring, such as vanilla, was added to the mix just before freezing.) Then the ice cream is pumped into containers. Containers of ice cream are stored in a huge freezer that is 20 degrees below zero. Here the ice cream becomes hard.

Stabilizers are important for storing hard ice cream. Ice crystals slowly grow larger when ice cream is stored in freezers. Stabilizers keep the tiny ice crystals tiny. They make sure that the ice cream is smooth even if it is eaten months later.

The ice-cream plant usually delivers its product to supermarkets and grocery stores and ice-cream parlors. Workmen dressed for winter load trucks that are freezers on wheels. The ice-cream plant sells ice cream to the stores. The stores sell it to you. When you eat it, you decide how much you like it. You decide if you will buy that kind again.

A scoop of ice cream doesn't cost you very much.

But your money goes a long way. The stores charge you a little bit more for your ice cream than they pay the factory. This small amount of extra money is their *profit.* The ice-cream plant uses the money it gets from the stores to pay its workers and the people who supply the ingredients. The ice-cream plant also has to pay to keep its machinery running and its trucks rolling. After it has paid all its expenses, it should have some profit left over. The sugar people, the flavoring people, and the farmers get some of their income by selling to the ice-cream plant. Some of your money goes to them. The money goes from you, the customer, to the retailer, to the manufacturer, to the suppliers. This chain of people makes up the ice-cream industry. Everyone in the ice-cream industry who makes a profit pays some of that profit to the government as taxes. Tax money is used to pay for roads and schools and water and other things everyone needs.

Suppose one day all the customers stopped buying ice cream forever. It probably wouldn't hurt the supermarkets and groceries too much. They have other things to sell. But ice-cream parlors would close. Ice-cream plants would close. Farmers would lose an important market. They would have to find other places to sell their milk. They would feel the loss in their pocketbooks. So would the flavoring people and the sugar people. A lot of workers would lose their jobs.

You eat ice cream because you like it. But when you buy and eat ice cream, you do other things as well. You use up the ice cream so that more must be made. This keeps the people in the ice-cream industry busy. Your money, along with the money of millions of other ice-cream eaters, helps a lot of people make a living. So when you buy ice cream and eat it so you need to buy more, you become a small part of what makes the world work. Yea!

An Ice-Cream Taste Test

How can you tell a good ice cream from a poor one? Some ice creams cost more than others. Better ice creams have more butterfat and fewer ingredients than less expensive ice cream. The cheapest ice creams have lots of substitutes for milk, cream, and natural flavoring.

Professional ice-cream testers are very serious about judging ice cream. Here are some of the things they look for:

– Color — Is it attractive? Does it have the color you expect for that flavor?

– Package — Is it clean, neat, and full? Does it protect the ice cream?

– Melting — The best ice creams melt quickly at room temperature. The melted ice cream is a creamy smooth liquid without bubbles.

– Body and texture — The best ice creams are firm but drip easily. They feel creamy and very smooth in the mouth. Better ice creams feel "chewier" than less expensive ice creams. Better ice cream has more butterfat than less expensive ice creams. It doesn't taste as cold. Inexpensive ice creams taste very cold and can give you a headache.

– Flavor — Vanilla ice cream should be pleasantly sweet. You should be able to taste the vanilla. The ice cream should not taste "cooked" like milk that has been boiled.

Of course, the proof is in the tasting. Ice-cream judges use a scorecard for vanilla ice cream. Do your own ice-cream test with friends. Buy different kinds of vanilla ice cream. Get premium ice cream and some inexpensive supermarket ice cream. Taste ice cream by the spoonful. Think about the taste of each mouthful. Use the ice-cream scorecard to rate each brand of ice cream. Take a sip of water to clear your mouth after each sample.

Ice Cream Scorecard

Flavor	**Yes**	**No**
Cooked	_____	_____
Too much vanilla	_____	_____
Too little vanilla	_____	_____
Too sweet	_____	_____
Not sweet enough	_____	_____
Too sour	_____	_____
Too salty	_____	_____
Body and Texture		
Coarse/icy	_____	_____
Crumbly	_____	_____

Fluffy	______	______
Gummy	______	______
Sandy	______	______

If you say "no" to everything, you've got great ice cream.

Make Your Own Ice Cream

You can make delicious ice cream at home if you have an ice-cream freezer. An ice-cream freezer lets you whip the ice cream as it freezes. Most people don't have one. If you don't, you can make "still-frozen" ice cream in the freezer section of your home refrigerator. It won't rate very well on the body and texture part of your scorecard but it will taste good. Here's how to do it:

Ingredients and Equipment

- small cup
- measuring cups and spoons
- 2 small mixing bowls
- mixing spoon
- rotary eggbeater (get adult help if you use an electric beater)
- 1/3 cup heavy cream
- 1/2 cup whole milk
- 1/4 cup sugar
- 1 teaspoon unflavored gelatin (from an envelope)
- very hot tap water
- 1 teaspoon vanilla extract
- a metal ice-cube tray with the divider removed
- plastic wrap or aluminum foil
- rubber spatula

Making Still-frozen Ice Cream

Put one of the bowls and the beater in the freezer to chill them.

Put the gelatin in the cup. Add a tablespoon of cold water to soften it. Then add two tablespoons of very hot tap water and stir until all the gelatin is dissolved. (Get a grown-up to help you add boiling water.) The gelatin is your stabilizer.

Stir a little of the milk into the gelatin. Then pour the mixture into one of the bowls. Put the rest of the milk, the sugar, and the vanilla in the bowl. Stir with a spoon until all the sugar is dissolved.

Turn the dial for the freezer compartment in your refrigerator so that it is as cold as possible.

Put the heavy cream into the bowl that was in the freezer. Beat it until it is stiff. This will take a few minutes. The cream stands in peaks when you lift out the beaters. Chilling the bowl and beaters ahead of time makes sure you get whipped cream, not butter.

Pour the milk-sugar mixture into the whipped cream and beat again. Then pour your ice-cream mix into the metal ice-cube tray. Cover it with plastic wrap. Put it into the freezer compartment.

Your ice cream will start to freeze around the edges first. This may start happening in twenty minutes. But it may take longer, depending on how cold your freezer is. When it starts to freeze, pour it back into the bowl and beat again. Then put it back in the

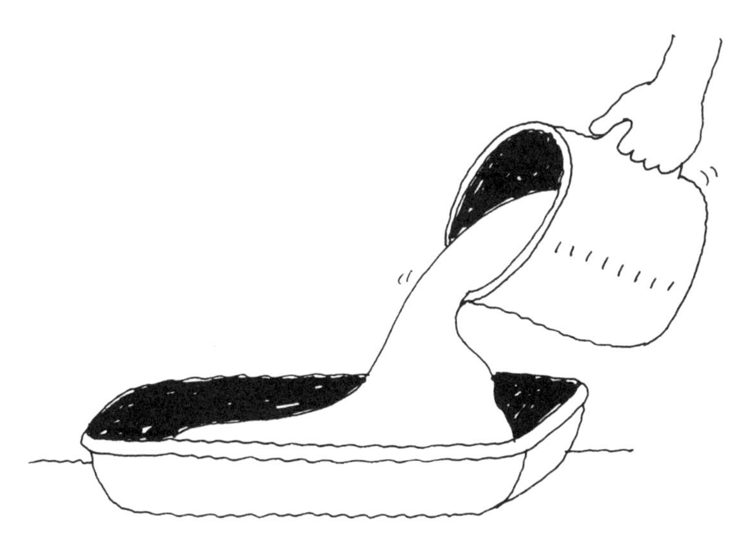

ice-cube tray. Use a rubber spatula to scrape out the tray and bowl so that you don't waste your ice cream. Beating keeps ice crystals small. When the ice cream starts freezing around the edges again, beat it once more.

Let your ice cream freeze solid before you eat it. This may take several hours.

Index

Africa, 21, 27
Air
  bacteria in, 10
  in ice cream, 42
Aluminum foil, 51
America, 24
Americas, 27

Bacteria
  cause disease, 38–39
  cause milk to spoil, 10
Baker's chocolate, 30
Beating ice-cream mix, 41–42
  home-made ice cream, 52–53
Body, of ice cream, 47, 48–49
Bourbon beans, 23, 24
Bowls, mixing, 51, 52, 53
Butterfat, 11
  in good ice cream, 47
  testing for, 34, 40

Cacao beans, processing, 28–30
Cacao trees, 26–28
Calves, 6, 7–8
Canes, sugar stored in, 14
Centrifuge, 18
Chocolate
  how chocolate is made, 29–32
  source of, 26–28
Chocolate factories, 29
Chocolate liquor, 30
Chocolate mix for ice cream, 36–37
Clam, 16
Cocoa, 31–32
  use in making ice cream, 33, 36, 37
Cocoa butter, 31
Color of ice cream, 47
Cooling milk, 10, 12, 38, 39
Cows, 6–7
  milking, 8–9
Cranes, 16
Cream, 11
  in ice cream, 35, 36, 51, 52
Cup, 51
Curing vanilla beans, 23–24

Dairy cows, 7
Dairy farmers, 7–12, 45
Delivery of ice cream, 43
Diseases, 38
Docks, 16

Eggbeater, rotary, 51, 52
Equipment for home-made ice cream, 50–51

Fat, in milk, 10–11
Feed for dairy cattle, 7
Fermenting cacao beans, 28–29
Filtering sugar, 19
Flavoring in ice cream, 42, 47, 48
Forklift trucks, 33
Freezing ice cream, 41–43
  beating while freezing, 41–42
  home-made ice cream, 52–53
Freighters, 16
Fruits, 42

Gelatin, 51, 52
Germs. *See* Bacteria
Granulated sugar, 19
Great Lakes, 16

Harvesting sugarcane, 14, 15
Holland, chocolate made in, 29
Hudson River, 16

Ice cream, qualities of good ice
  cream, 47–49
Ice-cream freezer, 42
  home freezer, 50, 52
Ice-cream industry
  chain of people in, 44–45
Ice-cream mix, 35–37
  ingredients for, 36–37
  pasteurizing, 40
Ice-cream plants, 11–12, 33–37, 39–45
Ice crystals, 41–43, 53
Ice-cube tray, 51, 52, 53
Impurities in sugar, 15, 19
Ingredients of home-made ice cream, 51
Inspection, government
  of dairy equipment, 39

Judging ice cream, 47–49

Laws on milk and milk products, 37, 38
Liquid sugar, 19, 20
  in ice-cream mix, 36
Liquors, 19

Madagascar, 23
Making ice cream at home, 51–53
Measuring cups and spoons, 51
Mexico, 21
Melting ice cream, 47–48
Milk, 6–12
  raw, 10–12
  use in making ice cream, 36, 51, 52
Milking cows, 8–9
Milking machines, 8–9
Milking parlor, 8–9
Milk solids, 34
  in ice-cream mix, 36–37
  testing, 40
Mingler, 18
Mix master, 36, 40
Molasses, 16, 18
"Mud," 15, 19

Nibs, chocolate, 30
Nuts, 42

Orchid plants
  vanilla beans from, 21–22

Packaging of ice cream, 47, 48
Pallets, 33, 34
Pasteur, Louis, 38
  pasteurization, 39–40
Percolation of vanilla liquid, 25
Pipes, 9–10
  washing, 11
Plants
  make sugar as food, 14
Plastic wrap, 51, 52
Profit, from ice cream sales, 44
Proteins in milk, 10

Raw milk, 10, 11–12
Raw sugar, 16, 18
  forms sugar crystals, 18–19
Refining process, 16–20
Refrigerator, 50–53

Roasting cacao beans, 30

Scorecard for ice cream, 48–49
Shipping
 cacao beans, 29
 raw sugar, 16–17
 vanilla beans, 24
Skim milk, 34, 35
Spatula, rubber, 51, 53
Spoon, mixing, 51, 52
Stabilizer, 36, 37, 42–43
 gelatin as, 52
Starch, stored by plants, 14
Still-frozen ice cream, making, 50–53
Stirrer, in milk storage tank, 11
Storage tanks for milk, 10, 11, 34–35
Storing ice cream, 42–43
Sugar
 how sugar is refined, 18–20
 in home-made ice cream, 51, 52
 sources of, 13–14
Sugar beets, 14
Sugarcane, 13
 processing, 14–16
Sugar crystals, 16
Sugar mill, 14, 15
Sugar refinery, 16–20
Sugar syrup, 18, 19
Sweating vanilla beans, 23

Transporting
 cocoa, 32
 milk, 11, 12
 sugar, 19–20
 vanilla, 26, 32
Tank trucks
 milk tankers, 11, 12, 34–35
 sugar tankers, 19, 20, 34
Taxes paid on profits, 44
Taste test of ice cream, 47–48
Teats, 8
 washing, 9
Testing
 ice-cream mix, 40
 milk, 34–35
 sugar, 34
Texture of ice cream, 47, 48–49
Tropics, 21, 26
Tuberculosis, 38

Udder (milk bag), 7–8
 washing, 9
Undulant fever, 38

Vacuum, 19
 sugar boiled in, 16, 19
Vanilla beans, 22
 curing, 23–24
 made into vanilla extract, 25
Vanilla extract
 in home-made ice cream, 51, 52
 making, 24–26
 storing and aging, 25
Vanilla flavoring, 24–26
 added to ice cream, 42

Warehouse, 17
Washing
 dairy cows, 9
 dairy equipment, 11, 39
 ice-cream equipment, 37
 milk tanks, 35, 39
Water, 7, 10, 11, 41–42, 51
White mix (ice cream), 36–37